AF835606

Seedling Barrier Protection from Deer and Elk Browse

Prepared for
Ministry of Forests
Silviculture Practices Branch

by
Ian Booth and John Henigman

February 1996

Canadian Cataloguing in Publication Data
Booth, Ian (Ian A.)
Seedling barrier protection from deer and elk browse

Includes bibliographical references: p.
ISBN 0-7726-2787-8

1. Browse – British Columbia. 2. Seedlings – Diseases and pests – Control – British Columbia. 3. Deer – Control – British Columbia. 4. Elk – Control – British Columbia. 5. Forest management – British Columbia. I. Henigman, John Frank, 1947– . II. British Columbia. Silviculture Practices Branch. III. Title.

SB994.D4B66 1996 634.9'6697357'09711 C96-960028-3

Acknowledgements

Many people and organizations have made important contributions to this paper. We would like to thank Martin Buchanan and John Guillemaud, Canadian Forest Products Ltd.; Chris Runnals and Doug Bacala, Squamish Forest District; Andy Wains and Dave Bain, Fletcher Challenge Canada; Jeff Fisher, International Forest Products Ltd.; Chuck van Hemmen and Carol Reagon, Port McNeill Forest District; Ron Diprose, Sunshine Coast Forest District and Mark Salzl and Del Williams, Queen Charlotte Islands Forest District, for providing us access and information on their seedling protection sites. As well, we would like to thank John Glennon, C.F.E Equipment Ltd. and Keith Windell, U.S.D.A. Forest Service, for providing recent tree shelter and seedling protector literature and costs.

The authors

Ian Booth was employed with the Forest Health Section, Silviculture Practices Branch, as a technician through the University of Victoria, Biology Co-operative Education Program. John Henigman is employed with the Forest Health Section as the Forest Health Forester.

Disclaimer

Table of contents

Table of contents

Figures

Tables

Introduction

This paper discusses the methods of detecting and assessing deer and elk browse problems, and describes possible methods of protecting seedlings from browse damage using physical protection devices. The publication is intended for use by foresters and forest technicians concerned with minimizing deer and elk browse damage to commercial trees.

Deer (*Odocoileus* spp.) and elk (*Cervus elaphus*) are species native to British Columbia's forests. Both species tend to graze primarily on herbaceous and woody species of the early serial stages of plant succession. Openings in the forest canopy, as a result of disturbances such as blowdown or logging, provide an ample supply of forage material and are therefore common grazing grounds for deer and elk throughout the year. When preferred forage stock becomes depleted, the browsing pressure on planted and naturally regenerated seedlings increases dramatically.

Damage to coniferous seedlings by deer and elk usually occurs in three ways:

- terminal and primary laterals are browsed off, eliminating or severely retarding seasonal height growth and, in some cases, the seedling is completely eaten;
- fine roots can be broken as deer or elk pull on seedlings during browsing (Schaap and DeYoe, 1986), with newly planted seedlings sometimes being pulled from the ground; and
- continuous browsing of the terminal stem, with the result of increased probability for the seedling to form a multiple stem, reducing the commercial value of the tree.

The effects of excessive browsing on plantation seedlings include:

- extending the time for seedlings to reach green-up or free growing height;
- a reduction in overall wood quality as over-browsed trees tend to be stunted with knotty stems; and
- a decrease in seedling survival making it difficult to attain target stocking density levels.

It is important to consider the potential for deer or elk browsing when preparing a silviculture prescription (SP). The cost of most seedling protection methods is high and this cost must be considered when selecting the species that will be planted. In some cases alternative strategies, such as prescribing a lower density of a susceptible species and protecting these seedlings, will guarantee a component of the species on the site at an acceptable cost.

Seedling protectors and deer browse

Does a browse problem exist?

The greatest commercial damage tends to occur in areas with large deer populations, when small seedlings of a highly preferred species are planted into areas that provide little else as alternative forage (e.g., such as clearcuts burned the previous autumn) (Nyberg and Janz 1990). Elk browsing damage is less common but can be very serious where the animals congregate in large numbers. Elk damage is more broadly focused on young trees than recently planted seedlings, and involves both browsing and trampling.

The following sections describe browse problem indicators as well as suggested methods of assessing and preventing browse damage.

Preferences of deer and elk for coniferous forage in the south coast of British Columbia

Deer damage is much more of a concern to silviculturists than elk, because deer are more numerous and widely distributed. In the few locations where elk concentrate seasonally, damage can be very severe. (Nyberg and Janz 1990).

Of the commercial seedlings planted, western redcedar (*Thuja plicata*) seedlings are usually the most heavily damaged by deer and elk. As well, heavy browse damage to Douglas-fir (*Pseudotsuga menziesii*) seedlings is a common occurrence in areas adjacent to winter ranges. Figure 1 describes the preferences of deer and elk for coniferous forage based on diet analysis and field observation.

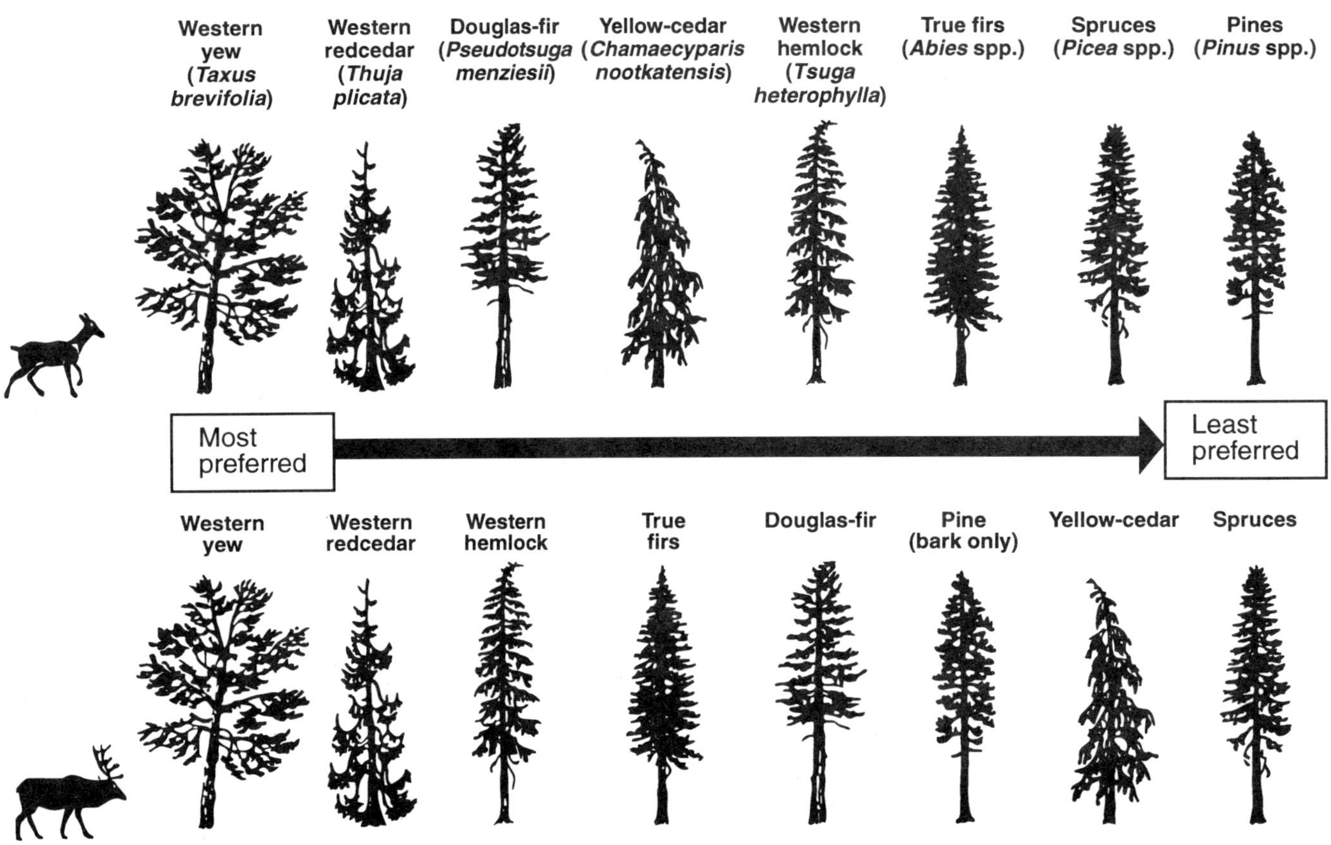

Figure 1. *Relative preferences of deer and elk on Vancouver Island for coniferous forage. [Taken from Nyberg and Janz, 1990] This sequence was derived from diet analysis (Gates 1968; Rochelle 1980) and field observations of Integrated Wildlife-Intensive Forestry Research (IWIFR) and other staff. Local variations in preferences can be expected.*

Seedling protectors and deer browse

Indicators for a potential browse problem (adapted from DeYoe *et al.* 1985)

Prior to surveying a plantation for seedling damage and mortality, or prior to conducting a Silviculture Prescription walkthrough, a number of questions should be asked:

- Are signs of deer or elk (hoof prints, faces, browsed twigs) readily seen in the area?
- Has browse damage occurred on previous plantations in the vicinity of the site?
- When does damage in the plantation, or those close by, occur? (In spring, at the time of bud burst?)
- Is the site surrounded by deer winter range habitat or does it appear to be located along a seasonal elk migratory route?
- Does the plantation provide a readily available and preferred forage source for deer?
- Are there signs of an established predator population in the area?

If the answer to one or many of these questions is yes, the potential for a deer or elk browse problem exists, and further action to evaluate and remedy any problem should be taken.

Important features of winter range habitat

Most browse damage will occur in planted cutblocks that lie adjacent to the winter range of deer or elk populations. Critical winter habitat is typically provided only by certain old growth forests. Old growth stands tend to provide an optimal balance of forage, security cover and thermal cover, all of which are features necessary for survival in the winter (Nyberg and Janz 1990). It is possible to identify winter ranges on topographical and forest cover maps. Winter range tends to be in or alongside old growth stands below 900 m elevation with gentle slopes and southern aspects. These stands will usually consist of tall, large-crowned conifers (providing 65%–75% crown closure) acting as thermal cover. As well, these sites usually contain a large shrub-understorey component providing forage material during the winter months.

In the lower mainland, cedar-hemlock thickets are common in these habitats as they provide security and additional protection from the cold.

These winter range site characteristics described are common for both black and white tailed deer and elk. Although the general requirements for deer winter range will remain the same, the flora found in each winter range will vary from one biogeoclimatic zone to another.

Survey for assessing seedling damage and mortality

A plantation survey is an excellent means of estimating seedling damage while providing vital information for future stand management decisions.

Sample plantation survey plan (adapted from DeYoe *et al.* 1985)
The procedure of this survey involves visually checking a quarter of the seedlings in a plantation and determining where browse damage is occurring and to what extent the seedlings have been damaged (Figure 2).

- For each seedling, in every fourth row, or in parallel transect lines 1 m in width and covering one quarter of the site, record a yes or no for browsing damage and survival.
- The tally sheet should have one column for each planted row of seedlings with the seedlings in each row represented by a number on the tally sheet.
- If possible, incorporate survey data onto a plantation map to determine if browse damage and mortality are uniformly distributed or clumped into particular areas of the plantation (Figure 2).
- Determine the percent of damaged seedlings by dividing the number of trees browsed by the total number of trees surveyed. Determine the percent survival using the same method.

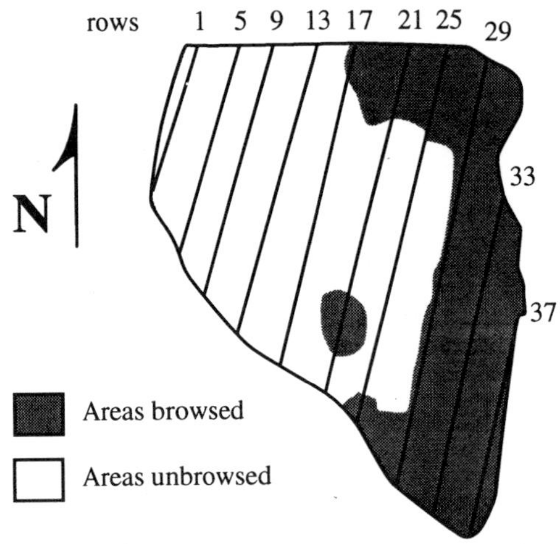

Figure 2. Sample map for seedling damage and mortality survey.

Questions to consider during the plantation survey

1. Does the extent of seedling damage and mortality warrant the cost of protection or can seedling browse and loss be tolerated because the remaining seedlings will meet target stocking needs?
2. Is browsing damage uniformly distributed, and if so, is it more cost effective to fence the entire plantation than to protect all or a portion of all individual seedlings?
3. Is the level of browsing damage and mortality so high in an area as to merit planting new seedlings to replace those killed while protecting surviving seedlings?
4. Does heavy plant competition seem to be partially responsible for areas of low browse damage with high seedling mortality?

Summary of seedling protection methods

This publication focuses on the physical methods of protecting seedlings from deer or elk browse. Other methods of browse protection including chemical, social or biological treatments, are briefly described below.

There are various types of chemical repellents which could deter grazers from seedlings. Contact taste repellents are products containing the fungicide thiram, applied directly to the plant. Systemic contact repellents, such as Ani-Pel®, offer a combination of chemical deterrent and fertilizer pressed into tablet form with slow release capabilities. When placed by the seedlings root system, the tablet deterrent and fertilizer move systematically into the foliage. The Ani-Pel® product has not yet become available for operational use. Area repellents repel by smell alone and are applied near or on the plant to be protected (Craven 1983). Examples of area repellents include plant extracts (e.g., garlic) and predator scents which mimic the odour of cougar, tiger or lion urine.

It is occasionally possible to encourage an increase in predator populations in heavily browsed areas to help reduce the numbers of deer. This method could be initiated at the time of forest site planning and should be done in consultation with wildlife managers.

Effective use of deer hunting is another approach for controlling deer populations and reducing browse pressure. For example, the hunting season is open 11 months of the year on Graham Island, Queen Charlotte Islands, with each hunter having a yearly limit of 10 deer. This is an exceptional situation, as the islands have an abnormally high deer population without any local predators. This option would require the full cooperation of wildlife managers and local public.

Physical methods for protecting individual seedlings from deer browse

The different classes of individual seedling protectors

'Tree shelters' form solid plastic tubes protecting the entire seedling from browse while providing greenhouse growing conditions at the base of the seedling (e.g., Tree-pro® and Tubex® tree shelters). Plastic mesh seedling protectors may form flexible (Tiller® netting) or rigid (Vexar® or Protex®) tubes designed to surround and protect a seedling from browse damage. Wire mesh such as stucco wire (18-gauge wire in a 2 inch square grid), may also be formed to act as tree protectors.

There are two basic functions of seedling protectors: those that protect the entire seedling and those that protect the terminal leader and some of the dominant lateral branches. Both plastic and wire mesh protectors may be used for terminal or full seedling protection, whereas, 'tree shelters' are designed for full seedling protection only.

Tree shelters and the greenhouse effect

A side effect of solid plastic tree shelters is the modified microclimate created around the seedlings which can improve establishment success and modify growth for many tree species.

Tree shelters trap carbon dioxide and moisture, while still allowing enough light for photosynthesis to pass through the shelter walls. This 'greenhouse effect' is created by driving the shelter into the ground, thereby sealing off the bottom end from air circulation. The shelters slows the dissipation of CO_2 given off by the soil, and moisture transpired by the plant. The resulting high humidity within the shelter helps reduce water stress, and the increased CO_2 levels can aid growth.

A three year study concluded that oak trees grow faster in height in tree shelters but do not differ in terms of total dry weight production (Windell, 1992). However, it is possible that tree shelter protected trees may not establish a strong enough root system during this time to support themselves once the protector is removed. The end result may be an unsupported tree prone to leaning that is still susceptible to browse damage. It is also uncertain if the warm, humid environment inside the tree shelter increases the risk of diseases normally found in nursery conditions (e.g., botrytis).

In areas where there are frequent freeze-thaw cycles in the winter, there is a risk of a snow or ice plug forming inside the tree shelter. This plug can crush or deform the seedling. To ameliorate this, some manufacturers have created ventilation holes in their tree shelters to reduce humidity in the tube and allow water drainage.

Physical methods for protecting individual seedlings from deer browse

Description of various available seedling protection products[1]

Tree-Pro®

This full seedling tree shelter is made of a single faced polyethylene sheet rolled into a tube and is designed to be attached to a support stake with lock ties. The flared rim of the Tree-Pro® protector is designed to prevent scraping of the trees bark as it emerges from the protector (Figure 3). This tree shelter design provides a greenhouse environment around the seedling while acting as a protector from deer, mouse, vole or rabbit browse (as claimed by the manufacturer). The manufacturer has recently started offering the product with ventilation holes, if requested, to reduce the humidity in the tube. Like other polyethylene products, Tree-Pro is designed to photodegrade after offering five years of protection. A newer product, Tree-Pro Junior® is designed to photodegrade in three years.

These tree shelters appear to offer excellent protection against browsing. The potential for nursery disease infection appears to exist with the original Tree-Pro® shelters as they create a humid environment at the base of the seedling. The fully enclosed design may lead to an increase in tree height without a corresponding diameter increase. Once the protectors are removed, the young tree may not have developed the root system and stem strength to hold itself upright.

Tree-Pro® tree shelters can be shipped flat on skid pallets, or can be shipped fully assembled as needed.

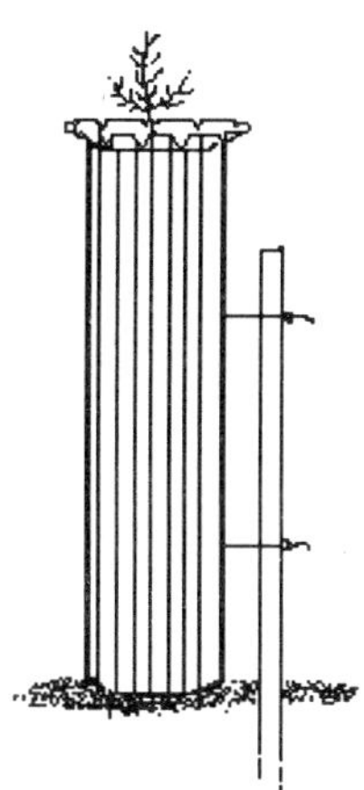

- Full seedling polypropylene tree shelter secured with stakes
- May be opened to allow trees to harden off
- Provides greenhouse atmosphere

Figure 3. Tree-Pro® tree shelter.

1 Names and addresses of manufacturers and distributors are provided in Appendix 1.

Physical methods for protecting individual seedlings from deer browse

Vexar® and Protex® tubing

Vexar® and Protex® polypropylene plastic netting formed into rigid tubes may be used as another form of seedling protection. The two products appear to perform similiarly in the field. Vexar® and Protex® tubing is designed as a full seedling protector to be used where severe damage occurs or is expected to occur by any species of wildlife (Figure 4). It appears that the tubing facilitates a small increase in the humidity surrounding the seedling and that this may be reducing the incidence of some seedling diseases such as sydowia tip dieback (*Sclerophoma* spp.). This disease often damages young drought or frost stressed conifer trees.

Tubing mesh patterns are of two types: diamond pattern resulting in a more rigid tubing, or twill-like pattern resulting in a more flexible tubing. A twill weave is one where the filling threads pass over and under two or more warp threads giving the appearance of diagonal lines. Vexar® and Protex® tubing are designed to breakdown when exposed to ultra-violet radiation. The manufacturer can offer tubing that photodegrades at different rates depending on the lighting conditions of the site.

Rigid seedling mesh protectors are normally less expensive than fully enclosed tree shelters. There appears to be a risk of protector induced seedling deformation with rigid seedling protectors. Regular maintenance is required to avoid this problem and to keep the leader protected. Seedlings can still be browsed as lateral branches grow through the protective mesh guard. This type of tubing does not appear to be as rigid as the tree shelters and the tubes may be susceptible to snow press.

Care must be taken to ship the rigid plastic mesh protectors so that they are not flattened. If flattened, time will be needed to reshape the tubes.

It is common to have to remove protectors that fail to photodegrade, in order to avoid tree damage.

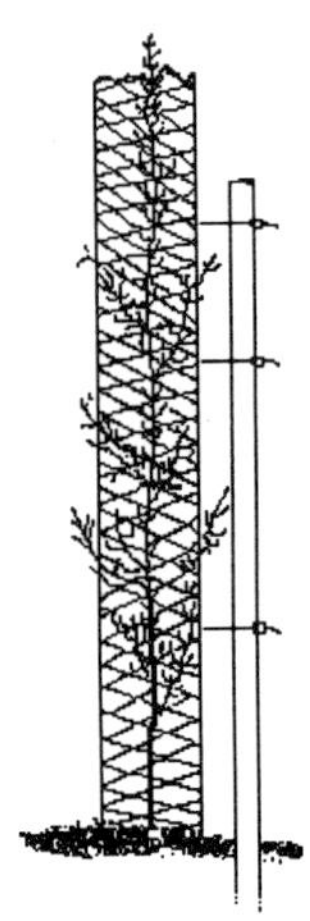

- Rigid polypropylene mesh tubing secured with metal or wooden stakes
- Rate of protector photo degradation can be selected depending on climatic conditions
- May be used as a full seedling or terminal leader protector

Figure 4. Vexar® and Protex® tubing.

Physical methods for protecting individual seedlings from deer browse

Tiller® netting

Tiller® netting is designed as a leader protector under light browse conditions and is not recommended for protection from vole or mice damage because of the risk of it being chewed apart. This light flexible plastic netting is designed to enclose the complete seedling or the top whorls of lateral branches and the terminal leader of the young tree (Figure 5). This netting is designed to fracture and degrade as the tree grows.

Although inexpensive, there appears to be drawbacks to using Tiller® stretch netting as a guard against deer or elk browse. According to many foresters consulted, installation is difficult. As well, it is difficult to install an adequate support system.

The manufacturer suggests anchoring the netting to bamboo or other stakes. Without support, the stretch netting may be blown off the seedling on windy, exposed sites. If the protector stays attached, seedlings can grow through the mesh. In this situation, the seedling will be left unprotected and tangled in its guard.

If the Tiller® netting was stretched open and held with a minimum of three support stakes, it may be possible for the seedling to grow up the centre of the guard, unobstructed and unbrowsed.

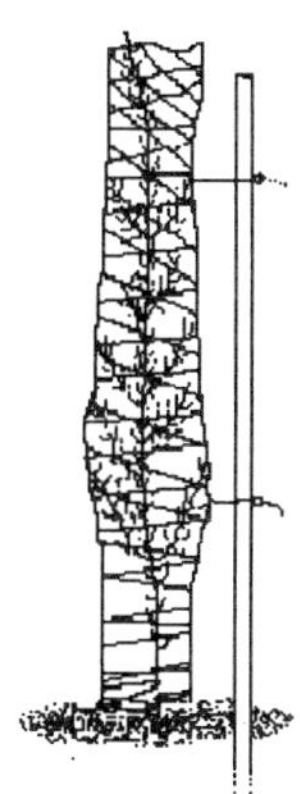

- Light flexible plastic mesh seedling protector
- Manufacturer recommends use for light browse conditions only as a leader or full seedling protector

Figure 5. Tiller® netting.

Physical methods for protecting individual seedlings from deer browse

Tubex® tree shelters

Tubex® tree shelters are seamless, twin-walled polypropylene tubes that photodegrade after approximately five years, or less, as designed. Built with a flared rim, to support the emerging trees without scraping the bark, these tree shelters are also installed with support stakes and ties to avoid being trampled or blown over (Figure 6). The manufacturer claims that the seamless molded design prevents scraping of the tender bark of young trees. Tubex® also claims to accelerate growth by maintaining a quasi-greenhouse atmosphere around the seedling. Tubex® tree shelters come in a variety of diameters and lengths to protect the seedling from different animals while accommodating different site characteristics and conditions.

The solid tube twin wall design of the Tubex® tree shelter appears to offer excellent protection against browsing damage from deer and elk however they appear unsuited for conifer growth (see the Queen Charlotte Islands Forest District sites in the section "Selected deer and elk browse protection project site descriptions"). If properly supported, the Tubex® tree shelter should be able to withstand the force of snow press.

The high humidity greenhouse atmosphere of the Tubex® shelter may make the seedling susceptible to infection by nursery diseases, such as botrytis. However, no specific problems have been reported to date.

The cost of shipping Tubex® tree shelters may be significant, as they must retain their tube shape during shipping.

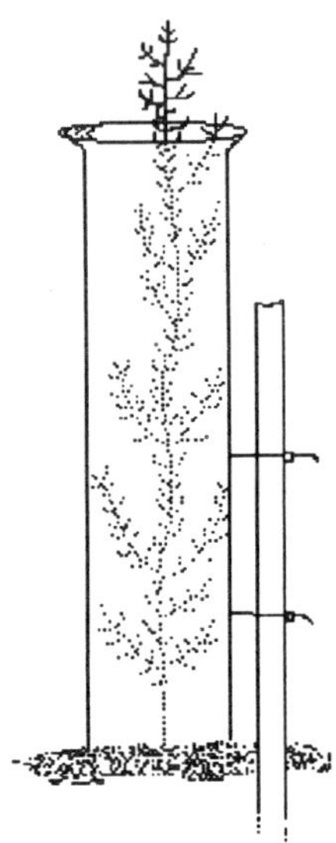

- Polypropylene tubes that photodegrade after five years or less, as designed
- Seamless design to prevent damage to young bark
- Tree shelter design provides greenhouse growing conditions for the seedling

Figure 6. *Tubex® tree shelter.*

Physical methods for protecting individual seedlings from deer browse

Stucco wire or square mesh wire individual seedling protectors

Stucco (18-gauge wire in 2 inch square grid) or square mesh wire can be used to create individual seedling protectors. These guards are made of rectangular pieces of mesh wire shaped into cylinders. Wooden stakes or galvanized metal pins are used to anchor and support the protectors while photodegradable lock-ties keep the cylinder closed and attached to the support system. Currently there are no commercial suppliers of wire protectors, the protectors must be custom made. The top ring of wire mesh may be clipped to make a 'thorny crown' further deterring deer or elk from reaching over and grazing on the seedlings.

After a few years in the field, wire protectors can become hidden from view. This situation can result in a worker hazard when brushing or spacing crews are using power saws.

Site visits to various deer damage control sites suggest that square mesh wire protectors offer the best browse protection without restricting seedling growth. When correctly installed, these large diameter guards appeared to decrease the possibility of protector induced deformities as the seedlings have plenty of room to grow without becoming tangled or woven into the guards mesh. When properly anchored, wire protectors appear strong and stable. Maintenance is still very important, and the mesh protector should be removed once the tree has reached a free growing stage to prevent tree damage and stand tending worker injury.

The cost of materials and shipping to the site can be high (see the section "Selected deer and elk browse protection project site descriptions"). The cost of equipment and time required to fabricate and install protectors must be considered. Properly constructed and installed wire protectors require little (though important) maintenance (i.e., usually only boosting to protect the leader at growing season 3 or 4). And, it may be possible to recycle these materials for use on other plantations.

Vandalism can be a problem. Some of the stucco wire protectors were stolen from Interfor's Malcolm Island trial sites.

Brush competition may be a problem as vegetation can grow inside the wire guards and avoid being browsed. The only place on Malcolm Island where salmonberry was observed was inside a wire seedling protector alongside a seedling (see the section "Selected deer and elk browse protection project site descriptions").

Whole seedling vs. leader protectors

Wind tends to cause a greater loss of terminal protectors than of total seedling protectors (Schaap and DeYoe 1986).

Iske (1989) reported that leader protectors did reduce browsing, but they had no significant benefit for height growth. Their use could not be justified as success could not be demonstrated.

Full seedling mesh protectors must be adjusted annually to cover the terminal shoot as it begins to emerge above the protector. If not, grazing will likely result in the loss of the terminal shoot.

There is an increased probability of deformation of the terminal leader branches when using full seedling mesh protection as branches may become 'woven' into the protector.

Under heavy browse conditions, full seedling protection is recommended as it has been shown that heavy grazing of the lateral stems can result in an increase in mortality rate or stunted growth.

Although the testing conditions vary significantly between trials, it is important to note that seedling protection relies entirely on selecting the protector best suited for the site characteristics (see the sections "Cost comparison for the various methods of seedling protection" and "Selected deer and elk browse protection project site descriptions").

Fencing and barrier methods available for plantation protection

Fencing methods

Fencing deer and elk out of a plantation can be a highly successful way of eliminating browse damage by ungulates. However, fencing methods cannot protect seedlings from other browsing animals such as mice or voles. Further, once a deer fence has been breached, the entire plantation is at risk. The materials and installation of deer fencing around plantations can be costly, though the per seedling cost is reduced with larger enclosures (see the section "Site characteristics to consider when selecting protection methods"). Installation costs will vary significantly depending on the condition, accessibility and terrain of the opening to be protected. Before committing to this type of plantation protection, a cost benefit analysis should be conducted taking into account the life of the fence, the materials required, and the person hours needed for construction and maintenance.

Deer fences can be classified into two categories: electric and non-electric.

Non-electric fencing

Fencing appears to be a practical alternative to individual seedling protection. For materials, the per seedling cost for fence protection will decrease as the size of the area to be protected increases (see the section "Cost comparison for the various methods of seedling protection"). However, installation costs will vary significantly from site to site depending on slope, soil characteristics, brush and slash present.

The main problem with fencing is that once breached, all seedlings are at risk to browse damage. Therefore, maintenance is crucial and will be a significant part of the treatment cost.

A wire mesh fence is effective if it is solidly constructed and at least 2.4 m (8 ft) high (Figure 7). Deer can jump over a 2 m (6 ft) fence with little difficulty (Craven, 1983). As well, the fencing should come close to the ground so that deer cannot crawl under it. In some cases slash may be piled at the base of the fence to act as another obstacle. Typically deer fences are constructed of two 1.2 m (4 ft) widths of page wire[2] attached to well supported 3.66 m (12 ft) posts. Other configurations include a 1.8 m (6 ft) fence comprised of one 1.2 m (4 ft) width of 5 cm × 5 cm (2 in × 2 in) square mesh nylon fencing above a 1.2 m (2 ft) skirt of chicken wire.[3] When piled with slash, the chicken wire skirt makes it extremely difficult for any animal to crawl underneath, while the smaller mesh size deters smaller animals from entering the plantation.

It is possible to save on materials by slanting the fence away from the protected crop. A slanted fence, constructed at an angle of 45° to the ground, only requires one 1.2 m (4 ft) width of wire. Deer will tend to walk under the overhang and be unable to jump over the wire strands. The fencing should be kept close to the ground to prevent deer from breaching the fence by crawling underneath. Using less materials, the slanting fence design may be significantly less expensive than the standard 2.4 m (8 ft) vertical fence described.

It may be possible to recover some of the fencing material for use on other plantations. Recovering and recycling these materials would significantly reduce the cost of protecting the next plantation.

---

2 Page wire, also called stock fencing material is a square grid mesh made of about 12-gauge wire.

3 Chicken wire is a hexagonal grid mesh made of about 20-gauge wire.

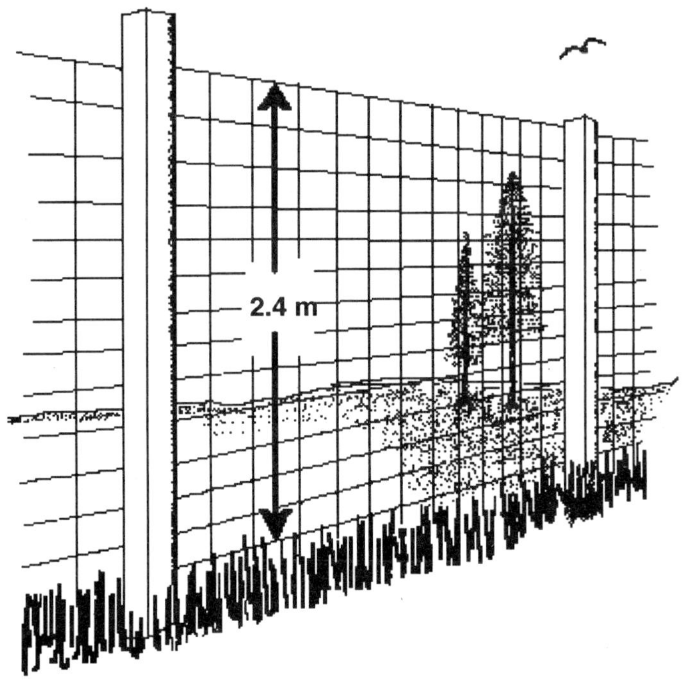

Figure 7. Non-electric deer fence.

- Made from two 1.2 m (4 ft) widths of page wire fencing joined one on top of the other and attached to 3.6 m (12 ft) fence posts.
- Total costs may be pro-rated over the life of the fence.

Fencing and barrier methods available for plantation protection

Electric fencing

Seedling protection with electric fencing was not observed by the authors. The main drawback to electric fencing would appear to be the maintenance required to ensure a shocking charge is deliverable to deter deer. Public encounters with the fence, may also pose a serious problem.

Both permanent and temporary electrical fence designs usually feature charged wires on outriggers forcing deer to encounter the hot wire before the fence (Figure 8). Generally, electrical fences require more upkeep than non-electric fences. Making sure the charger is working optimally and that batteries are running on full strength is necessary to ensure a strong deterring shocking charge. Downed vegetation, grounding the hot wires, must continuously be removed to keep a shocking charge high. As a precaution, a tandem charger arrangement will insure a constant power source if the primary charger malfunctions. Overall, electric fences require constant maintenance to ensure they are deterring browsing. For this reason it is unlikely that electric fencing is practical for B.C. commercial forestry.

There are a variety of electric fence designs available to fit different budgets, topography and browsing situation. Slanting electrical fence designs are also available and effective for protecting crops. Detailed information on the construction materials and methods required for other fencing designs may be found in Craven (1983).

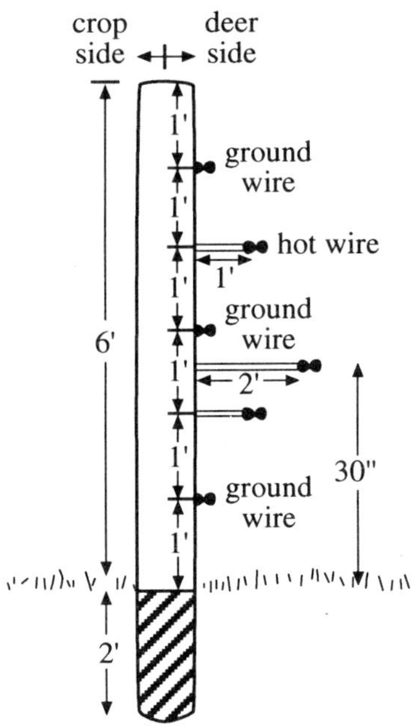

- Hot wires mounted on outrigger sticks so that deer encounter the hot wire before the fence
- Hot wires connected to charger terminal
- Tandem charger arrangement preferred to assure a constant power source if the primary charger malfunctions
- Ground wire connected to ground terminal which is in turn connected to a ground rod

Figure 8. Electric deer fencing.

Slash loading as a deer barrier

Large slash barriers have been used in attempts to protect plantations from grazing herds of deer or elk. As well, slash piles can be used in conjunction with fencing. Cowan (1945) has shown that deer dislike forcing their way through thick brush or heavy slash. However, it is difficult to rate the effectiveness and the cost of slash barriers for use in plantation protection. Informal trials have suggested that slash guarding is an unreliable means of plantation protection.

Site characteristics to be considered when selecting protection methods

Barrier toppling by wind

On windy sites, tree feeding barriers with inadequate support are frequently found leaning or toppled over. A leaning tree shelter will force the seedling to grow at an angle. A leaning, mesh style seedling guard will tend to force the seedling to grow through the mesh where it is susceptible to browse. Completely toppled protectors of any type will leave the exposed seedling unprotected. In anticipation of these problems, there are many ways of improving the stability and overall effectiveness of a seedling protector.

It is recommended that a minimum of two support pins, or one 'U' shaped pin be used per seedling guard. Metal pins are most frequently used to support rigid seedling protectors. They are woven into the mesh of the protector before being pounded into the ground. Cedar, pine, fir or cypress wooden stakes 2.5 cm × 2.5 cm × 1 m or 1.2 m long (1 in. × 1 in. × 3 ft or 4 ft long) attached to the seedling guard with photodegradable lock ties. This will provide excellent support against the forces of wind or snowcreep. The stake should penetrate the ground to a depth of at least 30 cm (12 in) and project about 1.2 m (4 ft) above the upper tie or wire, but remain below the rim of the shelter. Any object above the top of the protector rim could damage a seedling (Windell, 1992).

Staples are sometimes used to secure a protector to the support stake. The area around the staple where stress is concentrated, often breaks as the protector decays (Potter 1991).

Wooden stakes, purchased as factory seconds, cost approximately twice that of galvanized metal support pins. The overall cost of installation will be higher as wooden stakes are heavy and require tools for proper placement. This extra material and installation cost may be recovered in the long run as the plantation will require less maintenance and should have a higher rate of success. Recycle barrier supports if possible and, for safety reasons, please remove the metal pins.

Exposure of plastic barriers to sunlight

All plastic individual seedling protectors will become brittle and breakdown as they are weakened by ultra violet radiation. Depending on site aspect and exposure, rigid seedling protectors can be selected to degrade at different rates, depending on the period of seedling protection desired. A brush layer will delay the degradation of the protectors. The Vexar® tubing manufacturer established photodegrading rates according to tests carried out in Oregon state. The actual photo-exposure breakdown variable must be estimated before selecting a particular photodegrading rate of a rigid seedling protector.

Solid tube tree shelters allow only a portion of the available light through to the seedling, For example, opaque, coloured tree shelters can allow between 76 and 1 per cent of the available light through (Windell, 1992). This shading could slow tree growth.

Soil characteristics

Soil conditions may have an effect on protector selection and cost of installation. On rocky soil, erecting fence posts may be difficult. As well, different support systems for individual protectors might be required in rocky conditions. Wooden stakes or rebar are preferred over metal pins, which tend to bend when pounded into rocky soil.

Cost comparison for the various methods of seedling protection

Relative material, installation and maintenance costs for individual seedling protectors

The cost of seedling protection varies significantly between protection methods (Table 1). All estimates are for full size individual seedling protectors of similar dimensions. The material costs for the protectors will differ when used as leader protectors with different dimensions than the ones specified. All cost estimates are in 1995 dollars.

Table 1. *Per seedling cost estimates of available seedling protectors, protector supports and installation (based on 1995 prices for up to 2000 units)*

Protection method	Protector cost ($/seedling)	Acceptable support materials	Support cost ($/seedling)	Total protection materials cost ($/seedling)	Total installed cost[1]
Protex®[2] tubing 8.2 cm × 122 cm	$ 0.46	2 metal pins	$ 0.28 =	$ 0.74	$ 2.21
Stucco or square mesh wire[3] protector 45.7 cm × 122 cm tall	$ 3.63	2 metal pins	$ 0.50 =	$ 3.91	$ 5.38
Supertube®[4] tree shelter 122 cm tall	$ 2.36	2 metal pins	$ 0.28 =	$ 2.64	$ 4.11
Texguard® Treeshelter[5] 122 cm tall	$ 2.88	2 metal pins	$ 0.28 =	$ 3.16	$ 4.63
Tiller®[6] netting 122 cm tall	$ 0.26	1 metal pin	$ 0.14 =	$ 0.40	$ 1.87
Tree-Pro®[7] tree shelter 122 cm tall	$ 3.69	2 metal pins	$ 0.28 =	$ 3.97	$ 5.44
Vexar®[8] tubing 8.9 cm × 122 cm	$ 0.85	2 metal pins	$ 0.28 =	$ 1.13	$ 2.60

1 In each case, the planting and seedling protector installation cost was estimated to be $1.25 per seedling (average planting cost without the protector is about $0.48). This value was combined with the cost of a PSB 313A 1+0 plug seedling, estimated to cost $0.22, and added to the total protection materials cost to arrive at a total installed cost for each seedling and protector.

2 Protex® tubing is 14-strand rigid plastic and is priced at Vancouver, BC.

3 Wire mesh protector price is estimated based on discussions with forestry staff.

4 Supertube®, now available from Treessentials (section Appendix 1) replacing Tubex®, is priced in Canadian funds as available in St. Paul, MN, USA.

5 Texguard® Treeshelter is priced at Vancouver, BC.

6 Tiller® netting is available in 1524 m rolls and is priced at Edmonton, AB.

7 Tree Pro® is priced at Edmonton, AB.

8 Vexar® tubing is 14-strand plastic and is priced at Edmonton, AB.

Cost of protective fencing

A 2.4 m (8 ft) non-electric fence supported with 3.6 m (12 ft) pressure treated wooden fence posts, set 3 m (10 ft) out of the ground, costs $14.33 per running metre ($4.30 per running ft) for materials, and $11.00 per running metre ($3.30 per running ft) to install. When installed, this fence should have an effective life of 20 years. Therefore, it might be possible to recycle some of these materials, in particular the wire fencing, for use on other protection projects. It should be noted that hidden costs, for example fence gates for plantation access, have not been factored into these estimates.

With many different configurations, it is very difficult to provide a material, maintenance or installation cost estimate for electric fences.

Installation costs for both electric and non-electric fences will vary considerably with other factors such as transportation, plantation topography, soil depth and composition.

Cost comparison for the various methods of seedling protection

Estimated cost comparison for protecting 1- and 50-hectare plantations for six years

Table 2. *Estimated costs to plant and protect 1-ha and 50-ha plantations from deer or elk with various individual seedling protectors and with fencing (non-electric) for six years. Total costs are presented as net present values in 1995 dollars.*[1]

Protection method	Total protected seedling initial cost[2]	Annual per seedling maintenance cost	Cost to protect 1-ha plantation	Cost to protect 50-ha plantation
Fencing	variable[3]	$ 0.36	$ 10 359.00	$ 130 383.00
Protex® tubing	$ 2.21/2.18	$ 0 14	$ 2 728.00	$ 134 879.00
Stucco or square mesh wire protector	$ 5.38/5.03	$ 0.09	$ 5 713.00	$ 268 136.00
Supertube® tree shelter	$ 4.11/4.11	$ 0.12	$ 4 554.00	$ 227 682.00
Texguard® tree shelter	$ 4.63/3.91	$ 0.12	$ 5 074.00	$ 217 682.00
Tiller® netting	$ 1.87/1.84	$ 0.14	$ 2 388.00	$ 117 879.00
Tree-Pro® tree shelter	$ 5.44/3.42	$ 0.12	$ 5 884.00	$ 193 182.00
Vexar® tubing	$ 2.60/2.14	$ 0.14	$ 3 118.00	$ 130 879.00

1 Based on the assumption of a 1-ha enclosure and 1000 stems per ha. The number of individual protectors used for a site depends on the target stocking density for the site. Annual per seedling maintenance costs were calculated from information provided by staff of the Queen Charlotte Islands, Squamish and Sunshine Coast forest districts, and from Western Forest Products Ltd. The discount rate is set at 8%, and costs are presented in terms of the value of 1995 dollars.

2 Costs. The first unit price, transcribed from Table 2 is for a quantity of up to 2000 units, the second unit price is for a quantity of up to 50 000 units.

3 The installation cost for one metre of fencing is estimated at $22.57. The potential to reuse fencing material for up to 20 years has not been considered in this cost comparison.

Selected deer and elk browse protection project site descriptions

Squamish Forest District

Birkenhead River

This site lies in the transitional zone between the Coastal Western Hemlock and Interior Douglas-fir biogeoclimatic zones. Vexar® tubing, supported by two galvanized hooked metal pins, were installed to protect 16.8 ha of Douglas-fir (*Pseudotsuga menziesii*) and Western redcedar (*Thuja plicata*) PSB 313A stock. All seedlings on this site were planted and protected in 1990. The sites severe slope (average slope 65%) and snowcreep problem resulted in many protector tubes leaning, collapsing or buckling under the weight of the snow. Leaning tubes have resulted in a number of seedlings growing through the mesh of their protectors The cold weather appears to permanently deform the tubing as they tend to hold their new shape through the next growing season.

On this site, redcedar stock were observed spiraling within their tubes as they grew. Problems occur as one or more lateral or terminal branches grow through the protector mesh. As the seedling continues to grow, it 'pulls' the rest of the seedling to the same side of the protector. The rest of the seedling will then grow outside of the protector and will be browsed.

Blackwater Creek

This slightly sloping 27.1-ha opening is known to be used by a population of deer as winter range. Planted species on this site included both bareroot and plug western larch (*Larix occidentalis*), western white pine (*Pinus monticola*), western redcedar and yellow pine (*Pinus ponderosa*) seedlings. All seedlings in the block were planted and protected in 1991 with 12.5 cm diameter Vexar® tubing supported by two metal pins. The larch seedlings are growing successfully with Vexar® tubing when properly maintained. Redcedar seedlings have tended to spiral within the protectors, resulting in protector-induced deformities. Since the slope on this site is not severe, problems with leaning or toppled Vexar® from snowcreep or landslide were not serious.

Where both unprotected bareroot and plug seedlings were planted side by side, the plug seedlings were heavily browsed while the bareroot seedlings remained relatively untouched. This may possibly be due to differences in nutrient levels in the seedlings.

Port McNeill Forest District

Fletcher Challenge Site – Artlish River Valley

A total of 65 ha of productive valley-bottom was planted in 1990 with 2 + 0 western redcedar plugs. Several attempts were made to protect the seedlings from grazing by a resident herd of about 35 elk. Installed Vexar® tubing was being removed by elk, while chemical deterrents such as Ani-Pel®, failed to protect the seedlings from elk browse. Square wire mesh seedling protectors were installed and appeared to successfully protect the seedlings.

Material and installation cost for installing approximately 18 000 wire guards totaled less than $5.50 per guard. The wire mesh and wooden stakes were bought as factory seconds. Each protector consisted of a 1.3 m × 1.3 m square piece of 5.6 cm × 5.6 cm mesh rolled into a 46 cm diameter tube. Each tube was supported with a 5.6 cm × 5.6 cm × 1.3 m wooden stake, attached to the wire tube with lock ties. Guard induced seedling deformities were uncommon, and a very small percentage of protectors were bent and damaged by elk.

It is unclear whether these protectors can be reused. Their reuse would significantly reduce costs. Metal protectors or supports should be removed after their period of use to minimize the risk of injury to workers using brush or spacing power saws.

Selected deer and elk browse protection project site descriptions

Malcolm Island Forest Sites

Malcolm Island lies east of Port McNeill in Queen Charlotte Strait. It has a large deer population and no natural predators. Browsing intensity is very high. Most openings are classified in the Coast Western Hemlock (CWH) vm1 biogeoclimatic subzone and are blanketed with salal.

A. Interfor International Forest Products

Three Interfor sites were visited and are described as follows:

The site treatments were organized as trials. The first Interfor site visited was planted in 1987, with 313 western redcedar plugs, and protected with Vexar® protectors supported with two wire pins. On this site, deer were observed kicking and buckling the bottom of the installed tubes so that they could move their nose underneath and inside the protector. The deer were then observed pulling the seedlings outside of the guard where it could be browsed. Further, informal trials with three or more support pins did not appear to prevent deer from browsing in this aggressive manner.

The second site visited was also planted with 313 western redcedar stock and protected with rigid seedling protectors in 1987. This plantation was severely browsed and was not significantly re-stocked. During seedling protection trials on two occasions, 25 Vexar® protected seedlings were completely browsed within 24 hours of establishing the test plots. A trial was then conducted protecting western redcedar seedlings with 1.3 m high × 0.5 m diameter stucco wire protectors held together with photo-degradable lock ties and supported with two to four metal pins. The cost of the protectors was $3.75 per seedling. This stucco wire trial has been left unmaintained since its installation in 1987. To date, there has not been one recorded incidence of wire guard induced seedling deformity on this trial site. The cedar that grew to the top of the protectors and escaped further browsing are classed as free growing trees. Since the wire protectors were not maintained or boosted, some protected trees were browsed continuously as they emerged from the top of the wire guards.

The third trial site visited was planted and protected in 1991 with stucco wire protectors similar to those used in the second site. Here, the western redcedar were in their third growing season and appeared well formed and healthy while exhibiting excellent terminal elongation after each growing season.

Although expensive, supported wire guards offered excellent protection from deer browse and should be considered as a practical alternative for large scale plantation protection.

Interfor personnel suggested the following for protecting a site with wire guards. Plant to 60–70% of the target stocking density and protect with 1.3 m × 0.5 m diameter stucco wire seedling protectors adequately supported with wooden stakes and/or metal pins. Rely on naturally regenerated seedlings to meet target stocking densities. The following year, re-plant seedlings that died in protectors. Maintain the plantation once a year by servicing the wire guards and by reducing brush competition around wire guards. When required, usually after three or four growing seasons, boost the wire guards to cover and protect the terminal stem. In most cases the young tree will have enough foliage to hold its boosted cage without the need for additional support.

B. Pt. McNeill Forest District Tenure (forest cover mapsheet 92 L 066)

Four sites were visited with Forest Service personnel. Site characteristics are homogeneous across the island and ministry staff encountered the same site conditions and problems as did the Interfor personnel. All ministry sites visited were planted with PSB western redcedar (and in some cases yellow-cedar (*Chamaecyparis nootkatensis*)) 313B stock. All seedlings were then protected with Vexar® tubing supported with either straight 'J' or 'U' shaped metal pins. Two metre long metal wire bent as a 'U' appears to work best. Wire should be removed at the end of the project.

A seedling protection survey, conducted in 1987, noted that many of the protectors were being torn open by the deer. Many of the Vexar®tubes were found trampled. The single J and U-shaped pins appeared ineffective on this site in keeping the tubes supported against deer trampling. Stronger support stakes, made of wood or re-bar, are required to hold the protectors upright under these heavy browse conditions.

Selected deer and elk browse protection project site descriptions

Chilliwack Forest District

Canadian Forest Products Site #8662 at Chehalis Lake

This steeply sloped 11.5-ha site, located along the north east shore of Chehalis Lake, was originally planted in 1988, with PSB 313 Douglas-fir plugs to a density of 1300 stems/ha. A regeneration survey in 1990 indicated that deer browse had reduced stocking numbers on the site to a density of 472 stems/ha which included 20% naturally regenerated western hemlock stems. The site was replanted in 1992 with PSB 211 ½ + ½ Douglas-fir seedlings, each protected with 0.6 m × 12.5 cm Vexar® seedling protectors supported with two bamboo stakes.

Although snowpack conditions in this area are low, the winter of 1992 damaged many of the guards. A spring walkthrough in 1993 revealed many protectors, buckled under the weight of the snow, and holding their new shape because of the cold temperatures. As well, bamboo stakes did not appear to provide enough support against the creeping snow as many seedling protectors were found leaning or completely toppled over. Sliding rock and debris from roadside erosion also crushed many of the roadside installed Vexar® tubing. Annual maintenance of this site is required to correct problems that may have occurred over winter.

Queen Charlotte Islands Forest District Sites

Both the lowland and mountainous forest sites on Graham and Moresby Islands are highly susceptible to browsing from a large deer population. There are no predators and hunting pressure is of limited value in reducing deer numbers.

Opening 07 and 14 (forest cover mapsheet 103 F80)

This site was logged in 1988, and planted in fall 1991, with western redcedar 313B. These seedlings were then protected with 14 cm diameter Vexar® tubing supported with a galvanized metal pin. This single pin support appeared ineffective on this windy, exposed site as many of the seedling guards were found leaning or blown over. These seedlings were then open to browsing by deer. Trees supported by two metal support pins appeared more successful in protecting the guards from wind or animal damage.

As part of an informal species trial, yellow-cedar and Sitka spruce, (*Picea sitchensis*), were planted throughout this site. Both species, in particular the Sitka spruce seedlings, remained relatively undamaged as deer most frequently browsed the more preferred western redcedar seedlings or other surrounding vegetation.

Tubex® Tree Shelter Trial

In 1989, this site was planted with western redcedar seedlings and protected with 1.5 m (5 ft) tall Tubex® tree shelters. The Tubex® protected trees showed excellent height growth over the five growing seasons with many of these trees now 2 m in height. However, when the Tubex® was removed, the trees tended to lean over, under their own weight. Some studies show that, although Tubex® protected trees grow dramatically in height, there is not a significant increase in diameter accompanying this growth (Windell, 1992). It seems unlikely that these bowed trees will survive through their deformity. It may be possible to support these trees with stakes once the protectors are removed, however, this additional maintenance may not be cost efficient.

Sunshine Forest District

Texada Island Sites

During the summers of 1972 and 1973, the Ministry of Forests fenced in 6.4 hectares of a heavily browsed, 1969 Douglas-fir plantation on the south end of Texada Island. Fence posts were taken from unburned logging slash and were 15 to 25 cm in diameter and 3.6 to 4.9 m in length.

Post holes were dug at 4.5 m intervals and to a depth of 1 to 1.2 m, ground permitting. Corner posts required additional support with smaller diameter poles used as diagonal braces. The page wire used was in 15.2 m × 1.2 m (50 ft × 4 ft) rolls. Two, 1.2 m (4 ft) roll widths were used, one above the other, to provide an 2.4 m (8 ft) wire barrier. Logs and boulders were placed along the bottom of the fence to help prevent deer from crawling beneath it. Gates were incorporated every 30 m.

Deer fencing was removed from all blocks after the protected trees reached a free growing height.

It is now difficult to distinguish between the fence-protected trees and their unprotected neighbors. After 20 years there appears to be little detectable difference between the two stands, other than an apparent higher number of unprotected trees having multiple stems. A growth and yield analysis is needed to quantify the value of the fencing.

On a number of occasions, during the fencing trial, deer crawled under the fencing and were found browsing on the young trees. To prevent this from occurring, sites must be monitored continuously to ensure that the fence is not been breached.

The Sunshine Coast Forest District is currently monitoring about 30 operational Vexar® tube treated plantations on Texada Island. Problems identified, are similar to other Vexar® tube protected stands elsewhere on the coast, as described in this document. A high incidence of leader deformation has been noted if the Vexar® tubes were left unattended. As well, the Vexar® tubing does not seem to photodegrade as quickly as the manufacturer claims. In some cases, this results in the tubing affecting tree growth.

Seedling protection trials conducted using Tiller® netting were not successful. Without support, the stretch netting tended to act as a wind sock and was frequently blown off the seedling, on windy sites. Seedlings tended to grow through the mesh of netting. Many seedlings were seen unprotected or tangled in their guards.

The district staff have found that bamboo stakes rot too quickly and consequently, wire stakes are preferred. For safety reasons, wire stakes should be collected and recycled after use.

Literature cited

Cowen, I. Mc. T. 1945. The ecological relationships of food of the Columbian black-tailed deer (*Odocoileus hemionus columbianus* Richardson), in the coast forest region of southern Vancouver Island, British Columbia. Ecol. Monog. 15(2): 109–139.

Craven, S.R. 1983. Deer, published in: Prevention and Control of Wildlife Damage. Great Plains Agricultural Council Wildlife Resources Committee and Nebraska Cooperative Extension Service Institute of Agriculture and Natural Resources, Univ. of Nebraska-Lincoln. 10 pp.

DeYoe, D.R., deCalests, D.S. and W. Schaap 1985. Understanding and Controlling Deer Damage in Young Plantations. Oregon State University Extension Service Circular 1201. Corvaillis, OR: Forest Research Laboratory, School of Forestry, Oregon State Univ. 15 p.

Gates, B. R. 1968. Deer food production in certain seral stages of the coast forest. M.Sc. thesis. Univ. B.C., Vancouver, B.C.

Iske, K.W. 1989. *[Effectiveness and success of protection measure against game browsing in Norway spruce and Douglas-fir plantations.]* Wirksamkeit und Erfolg mon Schutzmassnahmen gegen Wildverbiss in Fichten – und Douglasienkulturen Allgemeine Forstzietschrift. No. 9-10, 245-246.

Miller, B.J. 1987. Measures to control deer browsing on Texada Island. Association of Registered Provincial Foresters of BC, professional report, F.I.T. 1485.

Nyberg, J.B. and Janz, D.W. 1990. Deer and elk habitats in coastal forests of southern British Columbia. B.C. Min. For. Research Branch, Victoria, B.C. Special Report series No. 5.

Potter, M.J. 1991. Tree shelters. Hdbk 7, Forestry Commission, London: HMSO.

Rochelle, J. A. 1980. Mature forests, litterfall and patterns of forage quality as factors in the nutrition of black-tailed deer on northern Vancouver Island. Ph.D. thesis. Univ. B.C., Vancouver, B.C.

Schaap, W. and DeYoe, D.R. 1986. Seedling protectors for preventing deer browse. Res. Bull. 54. Forest Research Laboratory, School of Forestry, Oregon State Univ. Corvaillis, OR. 12 p.

Windell, K. 1992. Tree shelters for seedling protection. USDA Forest Serv., Technology & Development Program. 9224-2834-MTDC.

Appendix 1 – Manufacturer and distributor addresses and telephone numbers

1. Vexar® tubing, Tiller® netting and Tree-Pro® tree shelter Canadian distributor:

   Canadian Forestry Equipment Ltd.
   17309 107th Avenue
   Edmonton, AB T5S 1E5
   Telephone (403) 484-6687, Fax 403-484-6763

   or

   Canadian Forestry Equipment Ltd.
   P.O. Box 31
   Vernon, BC V1T 6M1
   Telephone 1-800-667-4213, Fax 1-800-263-1878

2 Tree-Pro® tree shelters
   3180 West, 250, North
   West Lafayette, IN 47906, U.S.A.
   Telephone (317) 463-1011, Fax 317-463-3157
   1-800-875-8071

3 Supertube® Tree shelters

   Treessentials
   Riverview Station
   P.O. Box 7097
   St. Paul, MN 55107, U.S.A.
   Telephone (612) 228-0535, Fax 612-228-0554

4 Protex® Tubing Distributor

   Texguard Forestry Products Ltd.
   Box 139, Vanada, BC V0N 3K0
   Autofax/Phone (604) 486-7316